AF497693

Radisson's Journal Its Value in History

BY

HENRY COLIN CAMPBELL

[From Proceedings of the State Historical Society of Wisconsin 1895]

MADISON

STATE HISTORICAL SOCILTY OF WISCONSIN

1895

RADISSON'S JOURNAL ITS VALUE IN HISTORY.

BY HENRY COLIN CAMPBELL

[Address presented at the Forty-third Annual Meeting of the State Historical Society of
Wisconsin December 12, 1895]

Among all the subjects connected with the early history of
the Northwest, particularly that of Wisconsin it would be difficult to find one which is so deeply involved in doubt, confu-
sion, and error as are the careers of Pierre-Esprit Radisson and
Médard Chouart des Groseilliers

From a full belief in Radisson's Journal,[1] and in what has
been published concerning him, to a condition of skepticism on
many important points, has been a long and unpleasant road,
that I have traveled For, a year ago, when I began investi-
gating this subject, Radisson was to me one of the heroes of
our early history who seemed to deserve naught but honor
That vision has been gradually dispelled I still regard Radisson and Groseilliers as two of the most daring explorers who
penetrated the Western wilderness during the seventeenth
century, but I am convinced that Radisson, in his journal, is
guilty of gross exaggeration and downright falsehood in regard
to the exploration of the territory in and around Wisconsin.
He often allows his imagination to run riot In one place, for
instance, Radisson speaks of a little convention of three hundred
bears In another place he minutely describes a reptile that
nobody has ever seen on land or sea, a reptile that is absolutely
unknown to science[2] He calmly records the killing, during
one trip, of six hundred elk by himself, Groseilliers, and one
Indian He tells us moreover, of the shifting by the wind,
within a day, of fifty small sand-mountains from one side of

---

[1] See *Wis Hist Colls* , xi , p 64, for an account of the discovery and
publication of Radisson's Journal.

[2] See *Radisson's Voyages* (Prince Society Boston, 1885), p 69

Lake Superior to the other, the scene of this remarkable occurrence being not far from Sault Ste Marie   And to our still greater astonishment, he tells of sea-serpents in our great lakes Under the circumstances, I trust I may not seem too severe a critic when I accuse Radisson of drawing the long bow

Radisson's intentionally untruthful statements are almost matched by the unintentionally-untruthful statements regarding him and Groseilliers that have been made by some modern writers   Not very much has been written about these two men, but, in what has been written, the proportion of untruth to truth is surprisingly large   Error has been piled upon error, and hardly two accounts of any of the real or reported achievements of Radisson and Groseilliers agree

What is the historical value of Radisson s narrative of explorations in the West by himself and Groseilliers, soon after the middle of the seventeenth century?   The question is of the utmost importance, because it involves the discovery of the Upper Mississippi River, indeed it involves the first exploration of that great stream down to Southern climes,— for Radisson, in unmistakable terms, describes the Mississippi River, he states distinctly that he navigated its waters, and he asserts that he went southward so far that it never snowed nor froze All this took place if it did take place years before Joliet saw the West, years before Marquette reached America   Furthermore, there is every reason to believe that Radisson's narrative of the discovery and exploration of the Mississippi River was written several years before Joliet, accompanied by Marquette, embarked upon his famous voyage down that river, as far as the mouth of the Arkansas

Radisson was a mere youth when on May 21 1651 he arrived in New France   He was a native of St Malo, in Brittany, the place in which Jacques Cartier, the discoverer of New France, was born   Radisson's father was Sébastien Hayet-Radisson   and his mother was Madeleine Hérault [1] Both

----

[1] "I hope to embarke myselfe by ye helpe of God this fourth yeare" (meaning 1669), writes Radisson at the conclusion of his fourth voyage, speaking of Hudson's Bay   See his *Voyages*, p 245

[1] "Chouart et Radisson " by N E Dionne, in *Memoirs of Royal Society of Canada*, 1893 and 1894   The author is legislative librarian of the Province of Quebec

parents emigrated to New France, for Radisson states in his
Journal that they lived at Three Rivers Radisson had two
sisters, Marguerite and Françoise In 1646, Marguerite married Jean Veron de Grand-Menil, by whom she had three children Véron was killed near Three Rivers by the Iroquois,
August 19 1652, and a year and five days later his widow
married Groseilliers Françoise Radisson married Claude Vo-
lant de Saint-Claude and became the mother of eight children

Radisson himself, while he mentions in his Journal his
parents, his brother-in-law, and his brother-in-law s children,
never mentions having wife or child in New France, yet most
writers persist in giving him a family of his own in that
country There is no evidence that Radisson was married more
than once, and that was in after years to a daughter of John
Kirke,[1] one of the charter members of the Hudson's Bay Company To be sure, the registers of Three Rivers mention a
woman named Elizabeth Radisson, whose father's name was
Pierre-Esprit Radisson, his wife being Madeleine Hénault,[2] but
as our explorer was a mere youth when he reached New France
in 1651. and as Elizabeth Radisson married Claude Jutras,
called Lavallee, in 1657,[3] it is plain that she could not be the
daughter of our explorer, as some writers have stated It
appears that at that time there was another Pierre-Esprit Radisson at Three Rivers, and Dionne surmises that he was an uncle
of the younger Pierre Sulte,[4] writing several years before
Dionne, makes it appear that the elder Pierre-Esprit was the

---

[1] The Kirkes have been termed renegade French The fact is, that
Gervase Kirke, whose family had resided in North Derbyshire for several
generations was apprenticed to a London merchant, and in the course of
business became established for a while at Dieppe, where in 1596 he married Elizabeth Goudon David Kirke, who in 1628 attacked Quebec,
which surrendered the following year to his brothers Lewis and Thomas,
was a son of Gervase Kirke John Kirke, the father-in-law of Radisson,
was a descendant of David Kirke, and is generally designated as Sir John
Kirke, but he had not been knighted up to the time that the Hudson's
Bay Company was chartered by Charles II , for in that charter he is set
down as "John Kirke, Esquire

[2] Dionne, *Chouart et Radisson.*

[3] Benjamin Sulte, *Histoire des Canadiens-Français.* v.

[4] *Ibid*

father of the explorer, and that the former's widow married
Sébastien Hayet by whom she had three daughters, Marguerite,
Françoise, and Elizabeth   But as Marguerite was married for
the first time in 1646,[1] and as our explorer was not out of his
teens in 1651, he was undoubtedly younger than she therefore
Sulte's position cannot be supported   That there were at Three
Rivers two men named Pierre-Esprit Radisson, and that they
were not father and son, are made still more certain by the fact
that the parents of Elizabeth Radisson came from the parish of
Saint Nicholas-du-Chardonnet,[2] in Paris, whereas Sébastien
Hayet-Radisson and his family came from St Malo, in Brittany[3] It is certainly reasonable to suppose that the young
Pierre came from the same part of France that his father did

Groseilliers was considerably older than his dashing compan-
ion   As to the place of his birth, there is some dispute   The
Genealogical Dictionary of Canadian Families states that he was
a native of Charly-St -Cyr in Brie, a parish which cannot
now be located, but which may have been where now stands the
modern market-town of St Cyr-sur-Morin,[4] a short distance from
Meaux   Sulte states that the parents of Groseilliers lived at
Charly, parish of St Cyr, in Brie, but Dionne asserts that
Groseilliers was a native of Touraine, and in support of his position he quotes Mother Mary of the Incarnation   "Some time
since," the reverend mother wrote to her son in 1670, "a
Frenchman of our Touraine, named des Groseilliers, was married
in this country * * * He was very young when he came here,
and he cultivated my acquaintance because of our country, and
also in consideration of one of our mothers of Tours, with whose
father he had lived " Sulte says that before Groseilliers came
to America, he served at Tours in the family of Savonnière de
la Trouche whose daughter, Sister St Bernard, went to Canada with Mother Mary of the Incarnation   Groseilliers' service
in Tours would in far-away Canada entitle him to be called
"a Frenchman of our Touraine," and it does not follow from the

---

[1] Sulte and Dionne

[2] Dionne

[3] *Ibid*

[4] *Ibid*

reverend mother's letter, which Dionne quotes, that Groseilliers was a native of Touraine "the garden of France ' The *Genealogical Dictionary* and Sulte are probably right in stating that he was a native of Brie His father was Médard Chouart, and his mother Marie Poirier [1] There is no evidence to show that they accompanied their son to New France, where he arrived not later than 1641, perhaps as early as 1637 He too, was a youth when he arrived in New France He entered the service of the Jesuits in the capacity of donné,[2] or lay-helper and he remained with them for a number of years During this part of his career, he several times traversed the country between the French settlements and the villages of the Hurons, and in the course of his journeys acquired the Huron and Algonkin languages Sulte says that Groseilliers as early as 1645, went as far west as Lake Superior The next year he withdrew from the service of the Jesuits, and engaged in the fur-trade with the Hurons In November of the same year, he became engaged in marriage[3] to Marie Martin a daughter of Abraham Martin, a pioneer pilot of the St Lawrence, but instead of marrying her, Groseilliers, on September 3, 1647 became the husband of her sister Hélène, the childless widow of Claude Etienne It is an interesting fact, that Groseilliers' first wife was not only the daughter of the man whose name the historic Plains of Abraham bear to this day, but that she was a god-daughter of the great Champlain himself, who bestowed upon her the Christian name of his child-wife, Helene Boullé [4] By his first wife Groseilliers had two children, one of whom died the day that it was born while the other, bearing his father's name, has, like him, a place in history

While Radisson was generally known as Radisson, and by no other name, the man with whose fortunes his became linked was indifferently called Groseilliers and Chouart There is, in the whole province of Quebec, no land, no seignory, bearing the name of Groseilliers,[5] although Chouart is often called the Sieur

---

[1] Dionne

[2] *Ibid*

[3] Sulte

[4] Dionne

[5] Dionne, in a personal letter to the writer

des Groseilliers But by purchase, and by his marriage with the widow of Jean Veron, Groseilliers became possessed of considerable land in the vicinity of Three Rivers [1]

Radisson relates that early in the year (1652) following his arrival in New France, the Iroquois captured him while he was hunting near Three Rivers, and took him to one of their cantonments in what is now the State of New York   After one futile attempt to escape for which he was tortured and nearly killed, he succeeded in reaching Albany known at that time as Fort Orange He relates that at the fort he met a Jesuit who had been captured by the Iroquois, and that the Jesuit assisted him In the fall of 1653, Father Poncet, who had been captured by the Iroquois during the previous August, was at Fort Orange, and he relates a conversation that he had at that time and place with a young man who had been captured by the Mohawks at Three Rivers   There is no doubt that it was Radisson whom the priest met at Fort Orange, the latter's testimony is important, for not only does it corroborate Radisson's story about his captivity and his escape, but, combined with Radisson's statement that his capture by the Mohawks occurred the year after he reached Three Rivers, it proves conclusively that it was in 1651 that Radisson arrived in New France notwithstanding a statement by Sulte that he settled in New France before 1647

From Fort Orange Radisson went by way of Manhattan (now New York) to Holland thence to France, reappearing in May, 1654, at Three Rivers, where he had been given up for dead Upon reaching home he found that his sister Marguerite had, during the preceding August married Groseilliers   The friendship between Radisson and Groseilliers, who ever afterward were almost inseparable dates from that time their fortunes and their ambitions became one, they could not have been more firmly bound to each other had they been brothers in blood

Radisson calls his captivity among the Mohawks his "first voyage   Next in order and in number, in his published Journal is a voyage which he says he made, as part of the colonizing expedition and body-guard as well, which accompanied the

---

[1] Dionne

Jesuits Ragueneau and Dupéron to the Onondaga country, in the spring of 1657 This expedition returned to the French settlements after an almost miraculous escape from being massacred by the Iroquois, in April, 1658 [1]

Radisson next describes, in detail, two Western voyages which he and Groseilliers made after his return from the Onondaga mission The first voyage, Radisson says took three years, and during it Radisson claims that they explored the Mississippi River for a long distance The second Western voyage was along the south shore of Lake Superior, to the Huron village near the headwaters of the Black River, and to the Sioux Indians in Northern Minnesota Radisson says that this voyage included a trip to Hudson's Bay, and that it lasted two years

There is a conflict of opinion as to the route that Radisson and Groseilliers followed in coming West two French-Canadian writers [2] asserting that they ascended the St Lawrence to Lake Ontario, passed Niagara Falls, and navigated Lake Erie on their second voyage Radisson, however clearly intends to state that on both voyages he and Groseilliers went up the Ottawa River crossed Lake Nipissing, and descended French River to

---

[1] It is recorded, particularly by Mother Mary of the Incarnation, that a young Frenchman, who had been adopted by a famous Iroquois chief, told his Indian father that he had dreamt that he (the young Frenchman) would die unless a great feast was prepared and everything provided therefor eaten The chief loved his adopted son, and, to save his life, as he thought — for Indians are superstitious about dreams — he consented to the feast The Indians, encouraged by his son and by other French who were in the colony, so gorged themselves that they fell asleep, allowing the French to steal away in boats, which had been secretly built Dionne is positive that Radisson was the young hero of this interesting story But Radisson does not mention figuring as the youthful strategist, and there is no evidence that he was that young Frenchman It may have been Radisson, or it may have been some one else French captives among the Iroquois were not rare

[2] Dionne, in *Chouart et Radisson*, and L A Prud'homme, of St Boniface, Manitoba, in *Votes Historiques sur la vie de P E Radisson*, published in 1892

Lake Huron, the same route that Jean Nicolet followed when he visited Wisconsin in 1634 '

Describing the first voyage West, the "third voyage" of his Journal Radisson says that he and Groseilliers, with some of the Indians that had formed their party as far as the mouth of French River, went toward the South, and that while on this course they passed the place where the Jesuit fathers had lived meaning the destroyed missions among the Hurons, near the mouth of River Wye, Georgian Bay, and he virtually says that his party made almost a complete circuit of Lake Huron, "after * + * many days arriving at a "large island, where we found their [Huron' companions] village, their wives & children You must know that we passed a strait some 3 leagues beyond that place The wildmen give it a name it is another lake, but not so bigg as that we passed before We calle it the lake of the staring hairs, because those that live about it have their hair like a brush turned up ' Several writers, the late Edward D Neill[3] among the number, contend

---

' Radisson speaks of ascending the "river of the meadows, or crossing the ' lake of the castors, and of going down the 'river of the sorcerers to the 'first great lake Between the time that it was known as the Grand River of the Algonkins, the name which Champlain gave it and the time that it became known by its present name, the Ottawa River was called the River of the Prairies, as we learn in the Jesuit *Relations* In French, prairie is equivalent to meadow in English and in writing English Radisson used the term meadow ' The "lake of the castors is Lake Nipissing, which got the name that Radisson gave it, either from the fact that the Amikoue (beaver or castor) Indians dwelt not far from it, or from the abundance of beavers in the lake at one time Radisson s "river of the sorcerers, upon which he and Groseilliers descended to the "first great lake,' is French River, along which dwelt the Nipissing Indians who as the *Relations* inform us were called sorcerers because they practiced magic more than other Indians The first great lake" is of course Lake Huron See also, Butterfield, *Discovery of the Northwest in 1634*, p. 47

[2] Perrot and the Jesuit *Relations* lead one to believe that the Hurons, after fleeing from their own country in 1651 spent several years in the vicinity of Mackinac In 1653 they, or some of them, were at the Huron Islands, also called Pottawattomie Islands, at the mouth of Green Bay. The Hurons had certainly left Manitoulin before Radisson's first Western voyage

[3] *Wis. Hist Colls* , x , p 293

that Radisson's large island was Grand Manitoulin, Lake Huron
To me, this theory does not seem reasonable   Manitoulin Island
was out of the way of our two voyagers, according to Radisson's
description of the voyage   As they were coasting the east shore
of the present State of Michigan just before they reached the
"large island,' for them to go to Manitoulin would require a
voyage of about forty miles across open water   The   lake of
the staring hairs,' that Radisson mentions in his description of
the island, is certainly Lake Michigan, where dwelt the Ottawas,
who so dressed their hair that it stood erect   The strait which
Radisson mentions, to which "the wildmen give a name," seems
to be Michillimackinac, and he apparently intends to limit the
term to the narrow points of mainland between Southern and
Northern Michigan   Radisson's "large island" is undoubtedly
Bois Blanc, which has a shore line of thirty-five miles, and is a
few miles east of the two narrow points of the Michigan penin-
sulas   Bois Blanc would be on the way to the places which
Radisson says that he and Groseilliers afterward visited, while
Manitoulin island was not only out of their way, but to reach
it would necessitate a dangerous voyage across open water, and
the trip would have taken Radisson and Groseilliers back almost
to the place where they had entered Lake Huron before com-
mencing to skirt that body of water

From this large island where both Hurons and Ottawas seem
to have been at the time, Radisson says that he and Groseilliers.
not caring to stay upon an island, went to the Pottawattomies,
with whom they spent the winter, probably near Green Bay —
the bay not the city of that name   Radisson says that the following spring they visited the Escotecke (Fire Nation, also
called Maskoutens), who at that time dwelt upon Fox River [1]
That summer they, according to Radisson, explored Lake Michigan, "the delightfullest lake in the world " and thence went
upon their Southern journey   Radisson, continuing his narrative, speaks of visiting a country where the climate is so mild

---

[1] Nicolet found them on the Fox River in 1634, and Father Allouez, the
founder of the first Christian missions in Wisconsin found them in the
same neighborhood in 1670   Their village was apparently near Berlin, in
Green Lake County, see Thwaites's *Story of Wisconsin*, p 34

that the earth brings forth its fruit twice a year, so that "Italy
comes short of it," and of meeting people that dwelt about the
salt water (Gulf of Mexico) who told them of men who came
ashore in "great white things" (ships) [1] He also relates the
finding of a barrel, broken as they break barrels in Spain   Radisson continues   "We had not as yett seene the nation Nadoneceronons We had hurions wth us   Wee persuaded them to come
along to see their owne nation that fled there [the flight of the
Hurons to the Sioux on the Upper Mississippi River], but they
would not by any means '   Radisson speaks of seeing on this
journey the shovel-nosed fish, also a large bird, with a bill twenty-
two thumbs long, which swallows a whole salmon,— probably an
exaggerated description of the white pelican, which has a large
pouch under its bill, "shee-goats very bigg." probably antelopes,
"an animal somewhat less than a cow whose meat is exceedingly
good," perhaps wapiti and stags buffaloes, and turkeys He
describes "lemons not so bigg as ours, but sowrer," grapes
"very bigg, greene' — the vines grew by the river-side "It never
snows nor freezes there, but mighty hot "

Radisson and Groseilliers returned to the foot of Lake Michigan, visited the Indians of Sault Ste. Marie, and spent the following winter on the shore of Lake Superior, not far from the
Sault, in the midst of the nation of the Sault, who were Ojib-
ways, and in the neighborhood of Christinoes, or Crees The
question of the location of this winter camp is important, on
account of a journey that Radisson says that he and Groseilliers
made late that winter Radisson says that fearing the Iroquois
they retired to the upper lake, nearer the Nadoneceronons
This means that they went along the south shore, for had they
gone over to the north shore, they would have gone farther
from the Nadoneceronons, or Sioux, instead of nearer to them,

---

[1] I am not sure that Radisson does not go so far as to claim that he and
Groseilliers went clear to the Gulf of Mexico After leaving ' the delightfullest lake in the world " which is apparently Lake Michigan he says that
they went on until they found a climate superior to that of Italy and he
adds "Being about the great sea we conversed with people that dwelleth
about the salt water." The salt water is clearly the Gulf of Mexico, and it
seems that the "great sea is not Lake Michigan This is one of many
problems, that we find in the third voyage

as the Sioux were not far from Chequamegon Bay The fact
that they met Christinoes has given rise to a theory that they
went West almost to the Montreal River, on the south shore of
the lake, during this voyage but Radisson expressly says that
the Christinoes came to them in order to trade with the nation
of the Sault and to be where they could kill large game during
the winter   That Radisson, from his own account, did not go
very far west on the south shore of Lake Superior, is made apparent by the fact that in his second voyage West he minutely
describes the Pictured Rocks of Lake Superior and the adjacent
country, and intimates that he had never seen them before
That Christino camp was, therefore located somewhere on the
south shore of Lake Superior, between the Sault and the Pictured Rocks, possibly at Whitefish Bay

In connection with this journey to Lake Superior Radisson
makes a statement that is both surprising and confusing
Among the nation of the Sault, he says, "we found some
frenchmen y^t came up with us, who thanked us kindly for to
come & visit them " Early in his account of this voyage, Radisson states that the Frenchmen who had started with him and
Groseilliers from the French settlements, turned back, affrighted
by the Iroquois, leaving him and Groseilliers to continue the
voyage with no companions save Indians   Upon Radisson's
own showing, it is difficult to account for the presence of other
Frenchmen at the Sault   It is possible, of course, that some of
their original companions had afterward developed sufficient
courage to make a flying trip to the vicinity of the Sault, which
Nicolet had reached in 1634, and the Jesuits Raymbault and
Jogues in 1641   We have already seen that Groseilliers himself is credited by Sulte with a trip to Lake Superior in 1645

Late that winter, according to Radisson, he and Groseilliers,
with a hundred and fifty Indian companions, walked nearly fifty
leagues on snow-shoes, meaning one continuous journey for that
distance   They arrived at a river-side, where they stopped for
three weeks to make boats, and they then went up that river for
eight days, until they came to "a nation called Pontouatemick
& Matonenock, that is the scrattchers," here they obtained "some
Indian meale or corne from those 2 nations, wch lasted us till

we came to the first landing isle   There we weare well received again "   They tried to prevail upon the Indians of the "first landing isle" to take them down to the French settlements, but, the Indians being afraid of the Iroquois, Radisson says that he and Groseilliers were detained for another year   An incidental remark shows that these Indians were Hurons   " We weare in a great apprehension least that the Hurons should, as they have done often   when the flathers [Jesuits] weare in their country kill a flenchman "   The Hurons, after leaving the Mackinac and Green Bay regions, went to the Mississippi River country, and some years before 1660 were at Bald Island, Lake Pepin [1]   Is it possible that Radisson means that he found them there?   Did that journey of fifty leagues on snow-shoes, beginning at a point west of Sault Ste Marie, bring them to the mouth of Fox River?   Was it up that river that they traveled for eight days?   The Pontonatemck were Pottawattomies   Who were the Matonenock [2] Indians, whom they found with the Pottawattomies?   At this village they had to lay in a stock of corn meal   and this indicates that the journey from that point to the "first landing isle" was of considerable length   And the Hurons moreover, were the Indians whom they found at the "first landing isle '   From Radisson's description of the manner

---

[1] The movements of the Hurons are involved in considerable doubt   According to the Jesuit *Relations*, they were still in the Green Bay country in 1657   but we read that they lived for some years on Bald Island, Lake Pepin   We also know that Radisson and Groseilliers found them in Northwestern Wisconsin not later than 1659, and at that time the Hurons had been in that vicinity several months at least because the Hurons who went west with our two explorers knew the way to their village from Lake Superior, although the Hurons had gone to that place from the Mississippi River and had not yet reached Lake Superior   Radisson, speaking of their being at the 'first landing isle," says that they were " newly there "

[2] Radisson calls the Maskoutens the Escotecke, he probably does not mean the Maskoutens when he speaks of the Matonenocks   On a map attached to the Jesuit *Relation* of 1671, appears the name of the Mantououee, who lived near the Foxes at that time   In the *Relation* of 1673, they are designated as the Makoucoue   and they were still near the Foxes, in the Fox River region   At the time of Nicolet's visit in 1634 the Pottawattomies were near the mouth of Green Bay, and the Mantououee were near Escanaba

8

in which the "fiist landing isle" was ieached, it is simply impossible that it was Bois Blanc Manitoulin, or any other island
in that vicinity Besides, it is known that the Hurons were
not neai the stiaits of Mackinac noi farthei cast, at the time
that Radisson speaks of Radisson has before this intimated
that the Hurons had alieady gone up the Mississippi River

That jouiney which began with a tiamp of about fifty leagues
on snow-shoes was remarkable, if it actually took place, the
occasion foi it must have been extiaoidinaiy Radisson makes
it plain that the objective point was the place where the Hurons
dwelt and he has already said that he and Gioseilliers had pre-
viously endeavoied to picvail upon theii Huion companions on
that Southern tiip to visit the othei Huions in the country of
the Sioux I feel suie, from Radisson's account, that they weie
only a shoit distance fiom Sault Ste Marie when this journey
began They wanted the Hurons to escort them home, which
they wished to reach befoie anothei wintei set in Had the
Hurons been at Mackinac, or anywheie in that region Radissou
and Gioseilliers would not have had to start for their village
late in the wintei, nor would they have had to walk fifty
leagues on snow-shoes, spend eight days in ascending a rivei,
and go still faither, betoie reaching the dwellings of the Huions
The distance fiom Whitefish Bay to the mouth of ihe Fox River
is not much more than fifty leagues Oui exploiers were disappointed when they reached the "first landing isle," foi the
Indians iefused to take them down to the Fiench, and they had to
remain West anothei yeai, making thiee years altogethei That
summei, Radisson went hunting, and Groseilliers was attacked
with the falling sickness, or epilepsy They reached home,
Radisson says, the following year, "at last," he says, "we are
out of those lakes" The Indians with them, he states, numbered five hundied He adds that at the Long Sault, neai
the Ottawa Rivei, they were attacked by Iioquois, whom they
finally drove away, that aftei reaching Thiee Riveis he led an
onslaught against the Iioquois, whom he defeated, his foice con-
sisting of five hundred Indians and some Fienchmen and that
the Western Indians encounteied no enemy upon their ieturn
jouiney

Before summing up my conclusions regarding the 'third voyage,"— the first voyage West,— I find it necessary to take up the "fourth voyage " The reason therefor will be made apparent The route on this journey was up the Ottawa River across Lake Nipissing, and along the shores of Georgian Bay to Sault Ste Marie, where they rested and feasted Resuming their voyage they came to an isle "delightful for the diversity of its fruits," which they called the "isle of the four beggars," and the same night they went over to the mainland, a distance of about six leagues, and found themselves near the mouth of a small river, probably the Little Iron River, near which Radisson says he saw many pieces of copper He describes the Grand Portal, at the Pictured Rocks, and adds 'I gave it the name of the portall of St Peter, because my name is so-called, and that I was the first Christian that ever saw it " Radisson next describes the Huron Isles, and Keeweenaw Bay They portaged across Keweenaw Point and five days later they met a company of Christinoes At the mouth of the Montreal River, some of the Indians — Radisson intimates that they were of the nation of the Sault —left them to take the shortest route to their country, which was inland [1] At Chequamegon Bay, which Radisson describes with clearness, the Hurons who were of the party departed for the places where their wives were, "five great days' journeys" inland [2] It was cold, and Radisson says that he and

---

[1] Father Chrysostom Verwyst O S F , author of *Missionary Labors of Fathers Marquette, Menard, and Allouez*, and an excellent authority on the early history of Wisconsin, on the topography of the country south of Lake Superior, and on the Chippewa language and Indians thinks that the Indians who landed at the mouth of the Montreal River were Chippewas bound for Lac du Flambeau "Even to our day " says Father Verwyst, an old Indian trail led from Ironton Bay to Penokee Ridge and Lac du Flambeau

[2] Radisson, in describing his voyage to the same place, a few days later states that after traveling four days and just a day before they reached the Huron village, they reached a lake some eight leagues in circuit, ' which Father Verwyst thinks was Court Oreilles called Ottawa Lake by the Chippewas, even to this day there being a tradition among them that long ago Ottawas perished of starvation at this lake Radisson describes such a famine in that neighborhood, and Ottawas were among the victims

Groseilliers were nearly starved.[1] Near Whittlesey's Creek, or Shore's Landing[2] they built a small rude fort, the first structure built by white men in Wisconsin or on Lake Superior. Twelve days later fifty Hurons came and escorted the Frenchmen to their village. Soon the Hurons separated for the winter's hunting. They met again at a small lake and during the winter hundreds of them died of famine. Late in the winter, they wandered westward into the country of the Sioux, between the St. Croix River and the Upper Mississippi River, and in that country, between a small lake and a meadow the latter four leagues long a fort covering a space six hundred by six hundred and three feet was built — of course the first structure erected with the aid of white men in Minnesota. Radisson went three days' journey to the country of the Christinoes, and while returning to the fort he records that he passed a lake that was still frozen hard. At seven days' journey from the fort, Radisson and Groseilliers visited a village of the Sioux, or "nation of the beef" who claimed to number seven thousand men. After six weeks, the explorers returned to Lake Superior, accompanied by some of the Sioux, and found ice in Chequamegon Bay. They again built a fort, and afterwards, Radisson says, he and Groseilliers, accompanied by Christinoes, went to the waters of Hudson's Bay. Radisson says that they returned from the 'bay of the north' as he calls it, ' by another river." While returning, they received gifts from messengers sent by the Sioux, and in the middle of winter returned to the big fort which had been erected by them in Northern Minnesota. They returned home in the summer.

---

[1] There is a tradition among the Chippewa recorded by W. W. Warren in *Minn. Hist. Colls.*, v. that one morning early in winter two Frenchmen, the first white men to visit Chequamegon Bay were found in a starving condition on Madelaine Island. It has been surmised that these two men were Radisson and Groseilliers, and the surmise may be correct. But the tradition has it that these Frenchmen spent the winter in the Chippewa village on the mainland, while Radisson and Groseilliers spent the winter inland, with the Hurons and the Ottawas, and Radisson does not even mention being on Madelaine Island.

[2] Verwyst, 'Historic Sites on Chequamegon Bay.' *Wis. Hist. Colls.*, xiii, p. 433

In what year did this voyage end? There is a conflict of opinion on this point, but really there is no room for doubt The voyage of Radisson and Groseilliers to Lake Superior, to the Huron village in Northwestern Wisconsin, and to the Sioux in Northern Minnesota, terminated in August, 1660, although many writers claim that it was the voyage to the vicinity of Green Bay that terminated at that time

The Jesuit *Relation* for 1660 states, in brief that there arrived at Quebec, in August of that year, two Frenchmen, with three hundred Algonkins, in sixty canoes laden with furs, that the two Frenchmen had spent the previous winter on the shores of Lake Superior, that they had baptized two hundred children of the Algonkin tribe with whom they first lived, the children having suffered from disease and starvation, and forty of them dying, that the Frenchmen at six days' journey from Lake Superior, toward the southwest, found the remnants of the Petun tribe of Hurons, and that the daring explorers visited the country of the Sioux — Nadwechiwea, the *Relation* states meaning Nadouessioux,— among whom they saw women with their noses cut off, and round pieces of their scalps torn off, in punishment of adultery The *Relation* records that in five of these villages the two Frenchmen counted five thousand men

It is also stated in the *Relation* that the explorers went to the habitations of another nation called 'Bwalaks, or warriors,' who living in a country where timber was scarce, made fire with mineral coal, and covered their huts with skins, or made dwellings of clay Radisson it will be remembered, speaks of visiting the Huron village at five great days' journey from Lake Superior says that he and Groseilliers spent the winter with the Hurons and with a hundred and fifty Ottawa braves, who, with their families, joined them during the winter, and that before spring five hundred died of hunger He mentions finding in the Sioux country great cabins covered with skins and mats, and he records that the Sioux cut off noses, and removed the scalps at the crown, in punishment of adultery These Indians,[1] relates Radisson who calls them ' Nadonecero-

---

nons" and also 'nation of the beef,"—meaning buffalo,—had
no wood, and used moss for fuel

Radisson s statements and the account in the *Relation*, of
the two nameless Frenchmen who returned to Quebec from
Lake Superior in August, 1660, agree in almost every partic-
ular that is essential to the theory that Radisson and Groseilliers and the two nameless Frenchmen were identical Radisson, however says that they spent the first winter with the
Hurons, a considerable distance inland from Lake Superior
whereas the *Relation* states that they spent the winter on the
shore of the lake Radisson states that on this voyage he and
Groseilliers spent two winters near Lake Superior, the second
one at the large fort built in Northern Minnesota, but the *Relation* does not mention more than one winter that they spent
away from home on this voyage The *Relation*, moreover, does
not make the slightest allusion to the voyage to the Hudson's
Bay region, which Radisson asserts that he and Groseilliers
made while they were in the Lake Superior country

The *Relation* mentions the return of these two Frenchmen
from their Lake Superior voyage in August, 1660, but does not
give their names The following entry is found in the *Journal
des Jesuites*,[1] for 1660 "On the 17th [August] Monseigneur of
Petrea [Laval, the first bishop of Quebec] left upon his visit and
arrived at Montreal on the 21st, where the Ottawas had already
arrived on the 19th They numbered three hundred Des Groseilliers was in their company, who had gone to them the year
before They had departed from Lake Superior with one hundred canoes, forty turned back, and sixty arrived, loaded with
peltry to the value of 200 000 livres At Montreal they left to
the value of 50,000 livres and brought the rest to Three Rivers
They come in twenty-six days, but are two months in going
back Des Groseilliers wintered with the Boeuf tribe, who were
about 4,000, and belonged to the sedentary Nadoueseronons
Father Ménard,[2] Father Albanel and six other Frenchmen went

---

[1] *Journal des Jesuites*, par MM les Abbes Laverdiere et Casgrain
(Quebec, 1871)

[2] When the *Journal des Jésuites* says that Father Menard "went back
with them," it means that he went back with the Indians only But in

back with them   Albanel was soon abandoned by the Indians, and he returned to the settlements

Radisson himself furnishes conclusive evidence that the voyage which he and Groseilliers made to the Lake Superior country during which they visited the Hurons in Northwestern Wisconsin and the Sioux in Northern Minnesota, terminated in 1660 He records that, in returning from this voyage, his party passed the Long Sault, on the Ottawa River, shortly after the defeat of Dollard and his little band of heroes, one of the most thrilling and memorable events in early Canadian history   The massacre of Dollard's command occurred on May 21, 1660   Furthermore, speaking of passing along the south shore of Lake Superior, at the beginning of this voyage, Radisson clearly describes the Pictured Rocks near Munising, and he states that he called

---

Neill's chapter on " Discovery Along the Great Lakes " in Winsor's *Narrative and Critical History of America*, iv , p 170 in Winsor's *From Cartier to Frontenac*, and in other books too numerous to mention, we find the statement that Father Menard went back with Radisson and Groseilliers   An erroneous statement was never more widely circulated, upon such excellent authority   So far as I have been able to learn, Neill was originally responsible for it   Most of the writers who assert that Father Menard went west with our two explorers, imagine that it was the first Western voyage, from which Radisson and Groseilliers returned in August, 1660 but even admitting, for a moment, such an unwarrantable view of the matter, this theory that the Jesuit and the two explorers went West together is exploded by Radisson s own statement that he and Groseilliers rested for a year after their first Western voyage while Menard made haste to join the flotilla that had brought them home Some writers, with Radisson's year of rest in mind, start Father Menard and Radisson and Groseilliers West together as late as the autumn of 1661, regardless of the fact that Menard wrote his famous farewell letter, before starting on this voyage, on August 27 1660, at 2 o clock in the morning, that it is known that he started West at that time, and that he died in the wilds of Northwestern Wisconsin during August, 1661   Thus he had actually died before, according to these latter writers he started West   Winsor, to whom I took the liberty of writing when I saw the statement in his histories about Menard's coming West with our adventurers in 1660, replied in part as follows   I think you * * * may be right   I find in my interliued copy of my history (iv , p 170) that there is a ° against the passage " Verwyst the historian of Menard, in a personal letter to the writer, utterly discredits the theory that the priest and Radisson and Groseilliers came West together

what is now known as the Grand Portal, the "portal of St
Peter," because Peter was his name and because he was the
first Christian who ever saw it   Father Ménard, the first missionary to reach Lake Superior, passed the Pictured Rocks in
the autumn of 1660,[1] thus he not Radisson, would have been
the first Christian to see the Grand Portal, if those writers are
correct who assert that the second voyage, the one to Lake
Superior, did not end until after 1660   The dates that these
writers give, run all the way from late in 1662 to 1664   To
show how erroneous all these theories are, it is only necessary to
mention the fact that the *Journal des Jésuites* notes the presence of Groseilliers at Quebec in May, 1662 [2]

Did Radisson and Groseilliers really reach Hudson's Bay by
an inland voyage? Radisson says explicitly that they did so,
and it is one of the most important achievements claimed for
the two explorers   But the claim is a doubtful one   Radisson
says that this voyage to Lake Superior and beyond lasted two
years   It must have taken fully that time, if the two explorers
in addition to spending a winter anywhere near Lake Superior, and to visiting the Sioux in Northern Minnesota, made
a journey to the waters of Hudson's Bay   From reading the
Jesuit *Relations*, one gets the impression that the two adventurers spent but one winter in the West, and that impression is
strengthened by the *Journal des Jésuites*, which, in mentioning
the arrival of the Indian flotilla from Lake Superior in August
1660, states that Des Groseilliers was in their company, which
he had joined the year before ' It has been ascertained that
on April 15 1659, Pierre-Esprit Radisson was at Three Rivers,
as godfather of Marguerite, daughter of Groseilliers, Father
Ménard performing the ceremony [3] We have seen that there
were, then residing at Three Rivers, two men named Pierre-
Esprit Radisson, therefore, it cannot be stated with certainty

---

[1] Jesuit *Relations*, 1663

[2] Under May 1662, the following entry is found   'I departed from Quebek
on the 3rd for Three Rivers there met Groseilliers who was going to the
Sea of the North He left Quebek the night before with ten men '
During the same year Groseilliers and Radisson entered the service of
Boston merchants

[3] Sulte, *Histoire des Canadiens-Français*

that the godfather was our explorer, although it would have been natural for him to stand sponsor for his sister's child

The only contemporary writer who confirms the Hudson's Bay story in Radisson's Journal, is Noel Jérémie, who, in his *Relation of Hudson Bay*, where late in the seventeenth century he commanded for the French, states that Groseilliers had penetrated inland to Hudson's Bay and had also reached Manitoba Tending to confirm what Jérémie says, is the fact that, on at least one of the early French maps of the West what is now known as Pigeon River, at the Grand Portage on the north shore of Lake Superior bears the name of Groseilliers[1] Grand Portage is on the route to Hudson's Bay, and the fact that more than two hundred years ago Pigeon River bore the name of Groseilliers, indicates that our explorer had gone at least thus far, during his voyage which ended in 1660, for it is certain that he never visited that region after that year His presence at Grand Portage, at that time can only be accounted for by the theory of an attempt, at least, to reach Hudson's Bay from Lake Superior Radisson not only says that he and Groseilliers reached the "sea of the north,' as he calls it, but he speaks of barracks which he saw on the shore of the bay, barracks that Europeans had built, and he also states that the Indians of the bay told him that various white men had reached the place before, by water Radisson states that the journey to and from the bay was made in canoes and that the explorers returned from the bay on a different river from the one by which they went thither He says that they went direct from Lake Superior to Hudson's Bay, the statements that the Sioux sent gifts to them, by ambassadors " and that they spent part of the second winter at the large fort in Northern Minnesota, indicate that the more westerly route was that by which they returned from the bay

Radisson is at times most untruthful There is good reason, on this account alone, to doubt his Hudson's Bay story On the other hand he and Groseilliers seem to have started upon that voyage with the intention of trying to reach the "bay of the

---

[1] Franquelin's map, 1688 For descriptive and historical account of Grand Portage see *Wis Hist Colls* , xi pp 123–125, *note*

north," and we know that with both of them Hudson's Bay was a ruling passion

It being established that the second westward voyage of Radisson and Groseilliers terminated in August, 1660, this question assumes large proportions When did their first Western voyage, which has been assigned by most writers to the period actually covered by their last Western voyage, come to an end? The question is vital, because of Radisson s claim that his discovery of the Mississippi River took place during the first Western voyage

If Radisson and Groseilliers were not the two nameless Frenchmen mentioned in the *Relation* of 1656, who had spent the previous two years in the vicinity of Green Bay, I contend that the Mississippi River voyage which Radisson describes,— I mean the first Western voyage, from beginning to end,— never took place

In his account of the Lake Superior voyage Radisson speaks in several places of the other voyage to the West, and in so many words says that he and Groseilliers rested for a year from their first Western voyage, before they embarked upon their second Western voyage,— the one to Lake Superior, which was their last expedition to the West The two voyages are arranged in this order, in Radisson s Journal, the Lake Michigan voyage being called his third, and the Lake Superior voyage his fourth He could not declare more plainly, that the Lake Superior voyage was the next one after that to Lake Michigan In doing this, Radisson is caught in his own snare We have his own statement that he went to the Onondaga colony, accompanying the expedition which started in the spring of 1657, returning to Three Rivers in the spring of 1658 It is therefore plain that this Onondaga voyage took place between his two Western journeys, so that if the first Western voyage took place at all, it was undertaken at an earlier date than Radisson indirectly gives it

Radisson did not arrive in New France until 1651 One year later he was captured by the Iroquois, and did not return to Three Rivers from captivity until the spring of 1654 If Radisson ever made a Western voyage previous to his Lake

Superior journey, the earlier voyage took place some time between the spring of 1654, when he returned from captivity among the Iroquois, and the spring of 1657, when he went to the Onondaga country

At that period, the population of New France was so small,[1] that no two men — especially two men like Radisson and Groseilliers, one of whom had had a remarkable adventure with the Iroquois, and the other of whom was already looked upon as one of the most enterprising of explorers — could leave the French settlements for the far West, and return after a long absence, without attracting attention, especially on the part of the Jesuits, who, faithful chroniclers that they were, would of course have recorded what the explorers had seen and heard   Before this, Groseilliers himself had been for years in the service of the Jesuits   Hence I maintain that if Radisson and Groseilliers made a voyage to the West, between the spring of 1654 and the spring of 1657, they were the two nameless explorers of Lake Michigan and the Fox River country who are mentioned in the *Relation* for 1656   It may be asked why if Groseilliers was one of these two nameless French explorers, the Jesuits, his former masters did not mention his name in their *Relations*   With safety it can be asserted, in view of the small population of New France at that period, that no matter who those two explorers of 1654–56 were, the Jesuits knew their names, that some of the Jesuits even knew them personally and that they withheld their names for reasons of their own   It has been seen that the *Relation* for 1660 does not give the names of the two explorers of Lake Superior who returned in August of that year, but we know that they were Radisson and Groseilliers, because the *Journal des Jésuites* supplies the name of Groseilliers   The *Journal* was a more private record than the *Relations*[2] and was not published until 1871, while the *Relations*

---

[1] Garneau, in his *History of Canada*, says that even at a later period than this, the population of New France did not exceed 2,500

[2] For an account of the Jesuit *Relations* see Winsor, *Narrative and Critical Hist of Amer*, iv, also, *New England Magazine* for May, 1895   The only complete collection in America of the original *Relations* published in Paris is contained in the Lenox Library, New York

were sent to the court of France, and published soon after they were written

The apparent disagreement between Radisson and the Jesuit *Relations*, as to the duration of the Lake Superior voyage, has been noted   Radisson's assertion as to the time that his first Western voyage took place, and the statement of the Jesuit *Relations* as to the time that the two nameless explorers of 1654–56 spent in the West, differ in even a more pronounced manner Radisson, early in his account of this voyage,[1] says that it took three years, further on, he says that two years had gone by and that he and Groseilliers would not be able to return home for another year,[2] while near the conclusion, he says that the voyage had lasted three years and a few months[3] The *Relation* states that the two nameless explorers of 1654–56 started West on August 6, 1654, and returned toward the end of August, 1656 Radisson says that he started West with Groseilliers, on the first Western voyage about the middle of June (no year given), but a little further on, he contradicts this statement, for he says that, just before they reached Lake Nipissing, they picked some blackberries "not as yett full ripe," which they boiled with some *tripe de roche*[4] In the upper lake region, blackberries ripen about September 1   By July 1,— which if they started about the middle of June,[5] must have been about the time that Radisson and Groseilliers reached the spot where he says that he and Groseilliers picked the blackberries,— this fruit, instead of being nearly ripe would have been so green that nobody would think of using it for food If Radisson and Groseilliers had been the two nameless explorers who left the French settlements August 6, they would, when they reached the region of Lake Nipissing, have found blackberries in the state described by Radisson, for they would have reached that spot about August 20, at which time blackberries are nearly ripe   Radisson's statement about the blackberries disproves his statement that

---

[1] P 134

[2] P 157

[3] P 170

[4] A kind of lichen growing on rocks and used by early explorers as food

[5] This part of his journey took Father Allouez two weeks.

he and Groseilliers started for the West, on this voyage, about the middle of June, and it proves that if they did make such a voyage, they started at the same time that the two nameless Frenchmen did and that they were in fact identical with the latter

It is a significant fact, in this connection, that Radisson and Groseilliers cannot be accounted for at the French settlements during the period that the two nameless Frenchmen of the *Relations* were exploring the Lake Michigan region Radisson gives no account of himself between the spring of 1654, when he arrived home after his captivity, and the spring of 1657 when he joined the Onondaga colony On February 24, 1654, according to Sulte, Groseilliers was sergeant-major of the garrison of Three Rivers, and there is evidence of his presence at Three Rivers on September 29, 1656 Between these two dates, which is the period during which the two nameless Frenchmen were exploring Lake Michigan and the Fox River country, there is no record of the presence of Groseilliers at the French settlements on the St Lawrence

There are some striking points of resemblance between the experiences of the two nameless Frenchmen of 1654–56 and those described by Radisson in his account of his first Western voyage Both mention visits to the Pottawattomies and to the Maskoutens, both parties were disappointed by delay in returning home In both cases, mention is made of the joy which the return of the explorers caused, salvos of artillery being fired from the fort at Quebec Radisson says that the furs which he brought down on this voyage were a boon to the French colony, and, as a matter of fact, the condition of New France at that time was even worse than one would suppose from Radisson's words [1]

Between Radisson s tale and the Jesuit *Relations* there are

---

[1] Concerning the state of Canada in 1653, we read in the *Relations* that the keeper of the store at Montreal had not bought a beaver skin in a year that the Hurons kept away from Canada and that the Algonkin country was dispeopled The Quebec store house was empty And thus, the *Relations* state, "everybody has reason to be malcontent There is not wherewithal in the treasury to meet the claims upon it, or to supply public needs "

some points of difference almost equally striking The *Relations*, for instance, do not mention twenty-nine other Frenchmen starting westward and then turning back The nameless explorers told the Jesuits about the People of the Sea — the Puants, or Stinkards — our modern Winnebagoes, also, about the large nation of the Illinois, while Radisson who, if his account be true, must have seen both of these nations, says not a word about either of them Radisson mentions an encounter with the Iroquois, on the Ottawa while returning from this voyage, and he describes a battle that some Frenchmen and five hundred Indians under his command fought near Three Rivers with the Iroquois, whom they defeated As to both these events, the *Relations* are silent Radisson says that the Indians who went down to the French settlements with him and Groseilliers numbered five hundred, while the *Relations* state that two hundred and fifty Indians accompanied the two nameless explorers to the French settlements Radisson says that the Western Indians, in going back did not encounter the enemy, while we know from the *Relations* that the Indians who went to Quebec with the two nameless explorers were attacked by the Iroquois, and that Father Garreau, who with Father Druillettes, had been sent westward with the Indians, was mortally wounded, and the thirty Frenchmen in the party were obliged to return home

But, if Radisson and Groseilliers were the two nameless Frenchmen who explored Lake Michigan between 1654 and 1656, it is apparent that Radisson mixed fiction with facts, adding, for instance, fourteen months to the period of his voyage, hence, a few more falsehoods by him are not surprising

If Radisson and Groseilliers were not the two nameless explorers of 1654–56, that Western voyage which included the navigation of the Mississippi River never took place And even if they were the nameless explorers, Radisson's claim to the honor of discovering the Mississippi must be rejected for while it is possible that under these circumstances Radisson and Groseilliers did reach the Mississippi, the *Relations* contain no allusion to the fact nor is he supported oy any contemporaneous authority Radisson, who fraudulently extended the period of this voyage, if he did not invent the entire story, must have drawn upon his imagination for some of the territory that he claims to have explored, hence impeaches his own testimony

Why did Radisson lay claim to the discovery of the Mississippi? Certainly not to rob Joliet and Marquette of the honor, for Radisson's account of this voyage was written several years before Joliet and Marquette started upon their trip down that river Radisson and Groseilliers entered the service of Boston merchants during the year 1662 and in 1663 went in a Boston ship as far as Hudson's straits, the captain refusing to go any farther After litigation with Boston parties who violated a contract to furnish them with two ships for a voyage to Hudson s Bay,— a litigation in which our adventurers were unsuccessful,— they went to England at the solicitation of Col Robert Carr and Col George Carteret, two of the commissioners who in 1664 had taken possession of New York in the name of the British king It may be that Radisson's account of his first Western voyage was written in 1665, for the purpose of making an impression upon King Charles II or upon Prince Rupert, but it is certain that the journal of his fourth voyage was not finished in 1665 because at the end of it he describes the voyage of the ship "Eagle," in which, in 1668, he started for Hudson's Bay This vessel was forced by a terrible storm to put back, while Groseilliers, in the ship "Nonsuch," which started at the same time, continued on to Hudson's Bay It was the first voyage of our adventurers under the protection of England Radisson finished his report of his fourth voyage immediately after his vessel had been driven back to England

It appears to me that Radisson not only wanted the prestige of Western discovery in addition to the honor of discovery in extreme Northern latitudes, but he tried to impress the English with the desirability of acquiring possession of the fertile West, as well as of Hudson's Bay In speaking of his experiences in 1658, when he was about to make his escape from the Iroquois with the other French colonists in the Onondaga country, he says "It's sad to tend from such a place that is compassed with those great lakes that compose the Empire that can be named the greatest part of the knowne world ' Prophetic words, these.

The key note of his third voyage seems to be a desire to have
the English seize the region of the Great Lakes   It was not
until 1671 that the French formally took possession of the West,
and the suggestion of English seizure, was not altogether chi-
merical   Radisson's language, when he describes the far West,
is seductive [1] "The country was so pleasant, so beautifull &
fruitfull that it grieved me to see y^t ye world could not discover
such enticing countrys to live in   This I say because that
the Europeans fight for a rock in the sea against one another,
or for a sterill land or horrid country   * - *   Contrarywise,
those kingdoms are so delicious & under so temperat a climat,
plentifull of all things, the earth bringing forth its fruit twice
a yeare, the people live long & lusty & wise in their way
What conquest would that bee att litle or no cost, what
laborinth of Pleasure should millions of people have, instead
that millions complaine of misery & poverty! * * * It's
true, I confesse, that the accesse is difficult, but must say that
we are like the Coxcombs of Paris, when first they begin to have
wings, imagining that the larks will fall into their mouths
roasted, but we ought remember that vertue is not acquired
wthout labor & taking great paines   * * *   The further
we sojourned the delightfuller the land was to us   I can say
that [in] my lifetime I never saw a more incomparable country,
for all I have been in Italy, yett Italy comes short of it "

Radisson heard much about the Mississippi River, from the
Indians whom he met   He relates that an Iroquois chief told
him, during the voyage to the Onondaga country in 1657, that
he had once been captain of thirteen men who had gone against
the Nation of the Fire, and against the Staring Hairs, and on
this campaign had spent three winters away from home   Radisson says that the scene of the chief's story was in the
"upper Country of the Iroquoits, neer^a the great river that
divides itself in two "[2]   The Iroquois chief, according to Radis
son, told him of natives of that country who were of extraordinary height, two feet taller than he, and of tree fruit that is
"as big as the heart of an oriniack "   In his third voyage, Radis-

---

[1] *Radisson's Voyages* (Prince Society Boston), pp 150, 151
[2] P 106

son describes the Mississippi as the river that "divides itself
in two," and speaking of the "other river" he says   "These
were men of extraordinary height & biggness   *   *   *
They have fruit as bigg as the heart of an Orimack, wch grows
on vast trees wch in compasse are three armeful in compasse " [1]
The language attributed to the Iroquois chief, and that used by
Radisson, are suspiciously similar

I have never read anything more confusing than Radisson's
description of his third voyage [2]   It does not compare in clearness with any of his other narratives and the chief reason for
this is that Radisson has invented at least part of it

To sum up   The voyage of Radisson and Groseilliers to the
head of Lake Superior, and beyond, without doubt ended in
August, 1660   If Radisson's first Western voyage, the "third
voyage" of his Journal, took place at all, he and Groseilliers
were the two nameless Frenchmen who, during the period between 1654 and 1656, penetrated into the interior of Wisconsin,
by way of the Fox River, their voyage being almost identical
with that of Jean Nicolet in 1634   But even if Radisson and
Groseilliers were those two nameless explorers, the honor of
discovering the Mississippi River, which is claimed by Radisson,
cannot be bestowed upon them, because part of Radisson's
third voyage is clearly a fabrication so that, in effect, his own
unsupported testimony in regard to the discovery of the Mississippi is impeached by himself   "False in one thing false in
all "   Especially should this rule be applied to the statements
regarding the discovery of the Mississippi, an attractive enter-
prise which offered the strongest temptation to falsehood
Radisson's claim to the discovery of Hudson's Bay, by an inland route from Lake Superior has a stronger basis but even
that is in doubt

---

[1] P 168

[2] In justice to Radisson I have proceeded upon the theory that his ac
count of the third voyage is at least in part true   I have tried to locate
the places that he describes and to follow him in his wanderings, or in
what he says were his wanderings   But it is almost impossible to bring
order out of this chaos   The one point upon which I feel positive is that if
Radisson and Groseilliers were not the nameless explorers of 1654-56, the
third voyage described by Radisson never took place